A Guide to Radiation for the Everyday Scientist

Before you Start

This book assumes a general understanding of concepts covered in grade 10 science, mainly the components of an atom. If you need a quick refresher read the **Background Basics** section of the book before jumping in.

There are a wide variety of radiation-related topics in this guide; some topics are more science-heavy while others are less technical. The best way to read the book is each section in order; concepts sometimes reference past chapters. However, if you are particularly interested in one topic it is possible to understand the major components without reading all the previous material.

> **Side Notes** pop up in little chalkboards throughout the book. These are great for providing some helpful additional information.

Diagrams will show up throughout the book to illustrate concepts

This book will give you a good foundational understanding of radiation science and nuclear medicine. If, after you finish, you are interested in further learning, a great textbook is *Atoms, Radiation, and Radiation Protection* (1995) by James E. Turner.

Contents

- 1   Background Basics (optional)
- 4   What is Radiation?
- 5   Isotopes
- 9   Parent and Daughter Products
- 11   Radioactive vs. Irradiated
- 12   Radiation in Everyday Life
- 18   Types of Radiation
- 24   Dose Measurements
- 27   Interactions with Matter
- 30   Dangers of Radiation
- 36   Radiation Protection
- 41   Where Does Radiation Come From?
- 44   Decay Kinetics
- 47   How X-rays Work
- 48   Cool Uses for Radiation
- 51   A Brief History of Radiation

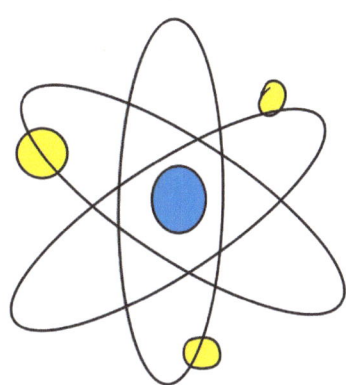

Background Basics (Optional)

Read this section if you want a quick refresher on the science of an atom.

Everything you see around you is made up of tiny building blocks called atoms. Atoms are too small to be seen even on a very powerful microscope. The atoms stick together to become molecules, which make up larger structures like cells or materials which then make up bigger objects you can see. Understanding atoms is an integral part of learning about radiation since most radioactive events take place at an atomic level. The starting point of any radioactive decay is an unstable atom.

The term "element" describes what type of atom we are talking about. If we think of atoms as small building blocks, then we can think of elements as the colour of the building block. Every element has a name and a 1-2 letter symbol.

There are 3 kinds of particles that make up atoms: **electrons, protons**, and **neutrons**. The different combinations of these 3 particles are what determines the kind of atom. This is talked about more in the "Isotopes" section of this book. For now, let's get an understanding of what these 3 particles are.

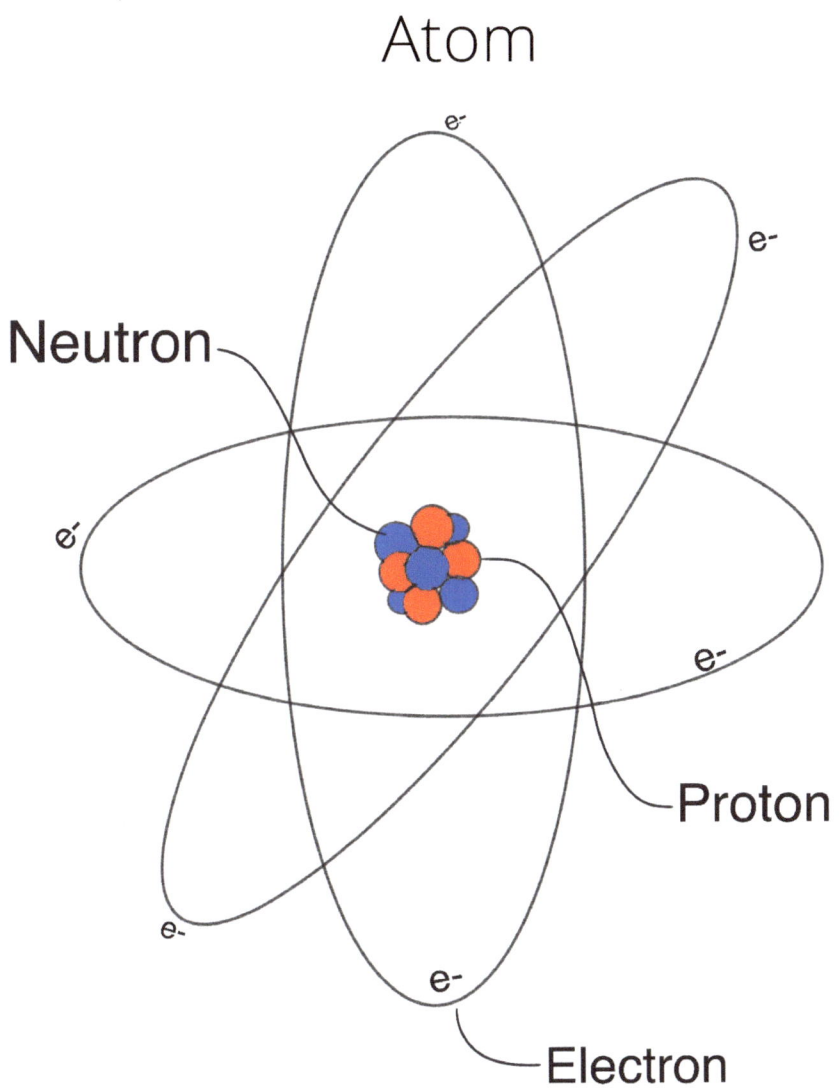

Protons are positively charged particles that are fairly large (as far as atoms go). The protons sit in the middle of the atom in an area called the **nucleus**. Since protons are charged, they repel one another. If you have ever held 2 magnets together with opposing ends you would have realized that they push back from one another. Now imagine a whole group of opposing charges stuck in a small tight room. That is what it is like for protons in the nucleus.

Neutrons are neutral particles that are the same size as protons. Neutrons also stay in the nucleus. The main job of the neutrons is to keep the protons evenly spaced. Even spacing is important to balance the repellant force of protons on themselves.

Electrons are the smallest of all 3 particle types. These particles weigh almost nothing and have a negative charge. The electrons float on the outskirts of the atom, orbiting around the nucleus. Their negative charge cancels the positive charge of the protons, making an atom neutral overall.

What is Radiation?

The term **radiation** refers to high-energy particles and electro-magnetic waves (the same waves that make up visible light). Natural radiation is emitted from atoms that are unstable. Each time the atom emits energy in the form of a particle or electromagnetic wave it becomes more stable. Radiation can be broken into two distinct categories: ionizing and non-ionizing.

Non-ionizing radiation does not have enough energy to cause damage and is not dangerous to humans. Some examples of non-ionizing radiation include visible light or radio-waves.

Ionizing radiation refers to radiation that has enough energy to cause damage to cells. Some examples of ionizing radiation are x-rays or alpha particles. This is what most people mean when they say "radiation" and it is the focus of this book.

Isotopes

Elements are defined by the number of protons they contain, also known as their **atomic number**. Any atom that has a specific number of protons will be the corresponding element.

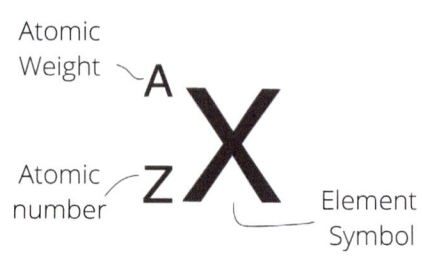

For example, an oxygen atom has 8 protons. Any atom that has more, or less than 8 protons is not oxygen. Similarly, carbon must have 6 protons, silver has 47, and uranium has 92. Each element has a set number of protons that define what it is. But atoms are made up of more than just protons, there are also electrons and neutrons to think about.

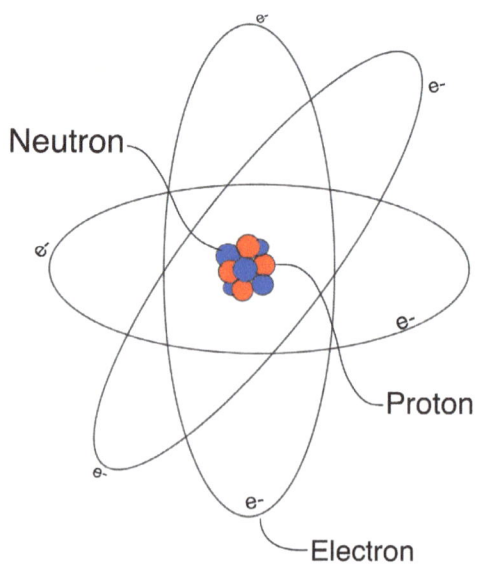

Isotopes tell us what version of a particular element an atom is. Neutrons are a variable quantity in an atom and they define an atomic isotope. Neutrons have the important job of keeping the protons in a nucleus separated and positioned correctly. In an average atom, the number of neutrons is generally equal to the number of protons. This is the most stable version of the atom for elements with low atomic numbers.

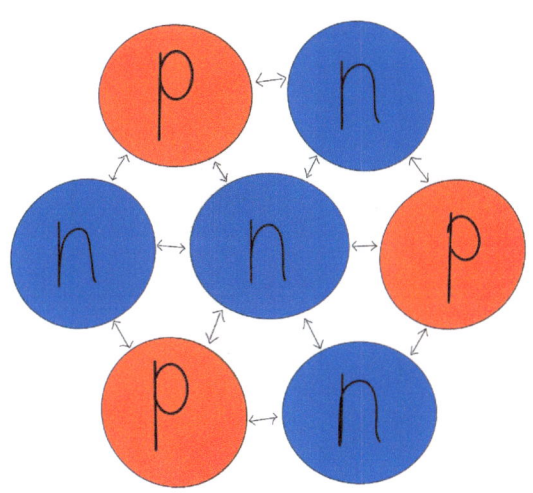

However, sometimes an atom has some extra neutrons or a few missing neutrons. This can cause the atom to be unstable since the forces present inside the nucleus are not well balanced. Atoms that are unstable can undergo radioactive decay to release energy and become more stable.

Since neutrons and protons are approximately the same weight, we define the isotope by the total weight of an atom. We write isotopes as the element name then the total atomic weight. Some isotopes are very common while others are extremely rare. We characterize this by **abundance** which tells us what percentage of that element is naturally found as each isotope version.

Helpful Hint: Think of an element like a selection of apples. There are different types, for simplicity let's say red or green. Each of these varieties is an isotope, a specific version of an apple. Abundance, therefore, tells us what fraction of our apple pile is made up of each type. Is it 30% green and 70% red or is the pile split evenly?

Example: Carbon-12 or C-12

Carbon-12 is the most common isotope of carbon and makes up 98.93% of all carbon on earth. In Carbon-12 there are 6 protons and 6 neutrons to make up a total weight of 12. The forces inside the atom are well balanced and it is extremely stable in this form. The other 1.07% of carbon is made up of Carbon-13 and Carbon-14. These isotopes are less stable and therefore less common. Carbon-14 is made up of 6 protons and 8 neutrons. This imbalance causes Carbon-14 to be so unstable that it undergoes radioactive decay.

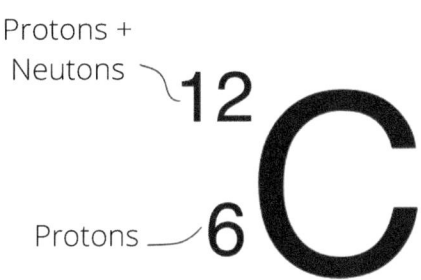

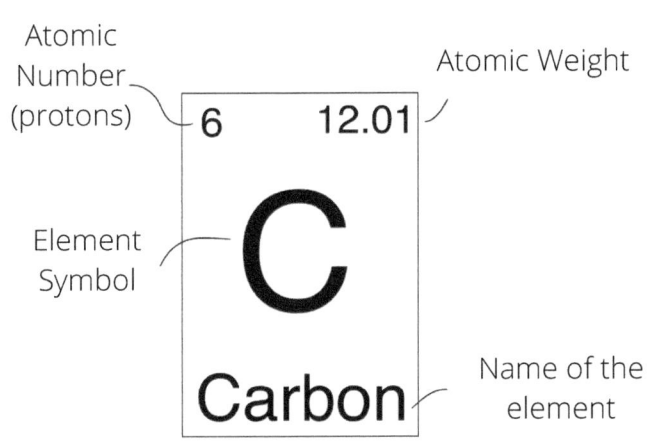

Parent and Daughter Products

When describing a radioactive decay situation, we use the terms parent and daughter to describe the isotope before and after decay. The **parent** isotope is the unstable atom we are starting with. This parent is radioactive and will undergo some form of decay by emitting particles or electromagnetic waves of radiation. The **daughter product** is what this atom turns into after it has released its radiation. Depending on what particles are emitted the type of element or weight may change. If a proton is ejected, the type of element changes because the atomic number is different. If a neutron is ejected, the weight of the atom changes and so the isotope number shifts down. In some cases, both protons and neutrons are ejected and so the element type and isotope weight change.

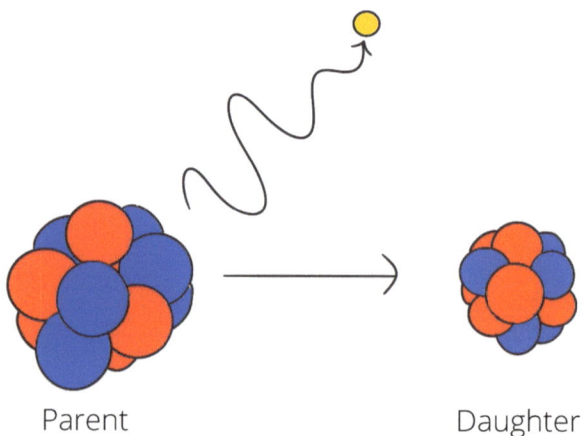

Parent　　　　　　　　　Daughter

It is important to note that a daughter product can be stable or unstable. If the daughter is unstable it will continue to decay and change until the atom lands on a stable configuration. This sequence of daughter products is called a **decay chain.**

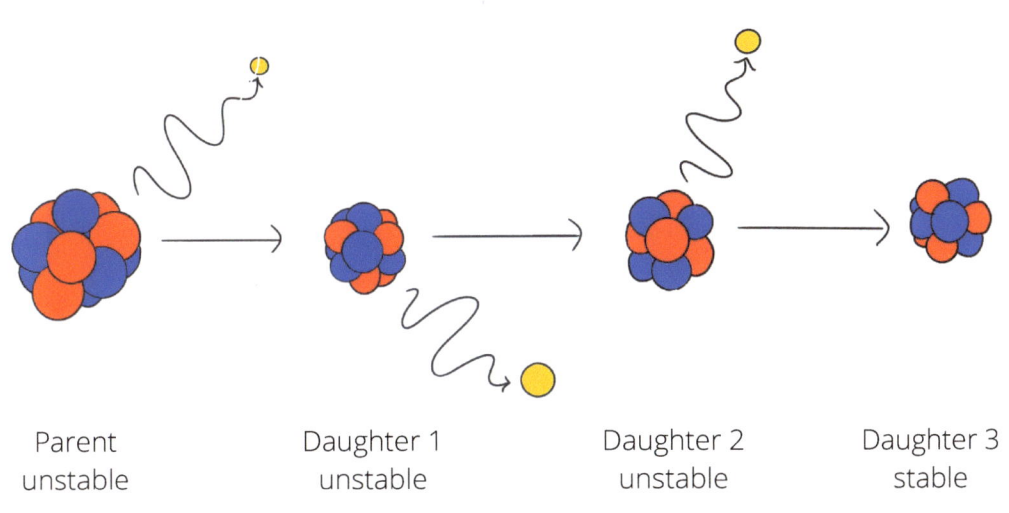

Parent
unstable

Daughter 1
unstable

Daughter 2
unstable

Daughter 3
stable

Radioactive vs. Irradiated

One of the most important distinctions to understand when learning about radiation is the difference between something that is **Radioactive** vs **Irradiated.**

Radioactive means that the item is emitting radiation. These items are considered a radiation **source**. Sources could include a radioactive element like uranium or a man-made device that creates radiation such as an x-ray tube. Naturally radioactive materials will continually emit radiation, but man-made devices can be turned on and off.

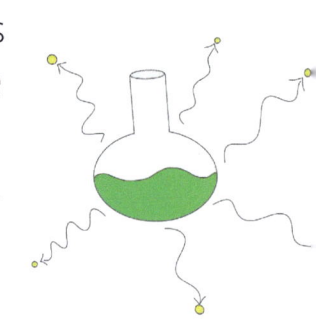

An **Irradiated** item is something that has been exposed to radiation. These items are not sources and do not emit any radiation. Once the radiation exposure stops the item is safe to handle and doesn't present any danger. A great example is when you get an x-ray taken. Your body is exposed to radiation (x-rays in this case) but once the machine is turned off, you are not radioactive.

Radiation in Everyday Life

Radiation is common in everyday life. Even though you may not realize it, you are exposed to small amounts of radiation all the time. This is known as **background radiation**. The dose of radiation is very low and causes no harm. The global average is 2.4 mSv per year (we will talk more about what these numbers and units mean later). Background radiation is caused by a wide range of contributors but some of the most common include radioactive elements in the ground (terrestrial radiation), cosmic radiation, and radioactive foods (ingested radiation).

Terrestrial radiation refers to radiation that is emitted from radioactive elements in the ground. The radiation from these different minerals and elements in the earth's crust is emitted most notably as electromagnetic waves. Of course, some of these elements are emitting radiation as sub-atomic particles but most of these are blocked and absorbed as they travel up to the surface.

Because emitting radiation can change the atoms' configuration (particularly in the case where sub-atomic particles are emitted) the element may change form. Materials that change to become a gas can seep out of the ground and are present in our atmosphere, where we breathe in the radioactive particles. The most common example of this is Radon-222 gas.

The amount of terrestrial radiation depends on geographic location and the natural material deposits in an area. A region that has many deposits of radioactive elements will have a much higher background radiation dose than a region with almost no radioactive elements in its soil.

Cosmic radiation refers to radiation coming from space. Most of this radiation is produced by stars, such as the sun, during their fusion and solar flares. The emitted particles interact with elements in the atmosphere as they travel towards the earth's surface.

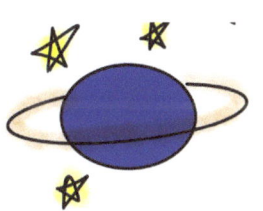

Instead of being based on geography, the dose from cosmic radiation relies on altitude. Since the atmosphere filters out a large portion of the incoming cosmic radiation, locations at high altitudes will receive a larger dose of cosmic radiation. Locations at lower altitudes have more atmosphere above them filtering the radiation and therefore receive lower doses of cosmic radiation. Flying in an airplane actually exposes passengers to small amounts of radiation while in the air.

Ingested radiation refers to radiation dose from eating foods with small amounts of radioactive elements. Food accounts for 1/8th of the natural background radiation people receive. Foods can acquire these radioactive elements in 3 ways: uptake, deposition, or bioaccumulation.

Uptake occurs when plant roots suck up radionuclides from the soil. One of the most common occurrences of uptake is from radioactive potassium. Potassium-40 is an isotope of potassium that is unstable and therefore undergoes radioactive decay. Potassium-40 makes up 0.012% of all potassium, so only very small amounts can be found in food. Potassium is found in foods such as bananas, lima beans, nuts, coffee, milk, and many others. Our bodies need potassium to function so avoiding the element all together is not a good idea.

Deposition happens when airborne radioactive particles settle on crops. This is a concern particularly in the case of a nuclear disaster such as the Chernobyl reactor meltdown of 1986. Crops in areas of nuclear fallout have to be destroyed to prevent ingestion of radioactive elements beyond safe limits. Ordinary crops do not have significant levels of radioactive deposition and are not a major concern.

Beyond just plants, meats and animal products can contribute to ingested radiation. In **bioaccumulation** radionuclides accumulate in animals that have ingested them from food or water. These elements are used in their bodies and can show up in muscle tissue, bones, milk, eggs, and more.

The previous examples have talked about natural radiation that shows up in our everyday lives. Most of this background radiation is unavoidable and has always been a part of human life. However, in recent years a new radiation source has begun to show up in our everyday lives in the form of medical procedures.
Medical Radiation now accounts for the majority of radiation exposure in the general public. A wide variety of procedures and diagnostic tests make use of different radiation forms to allow doctors to diagnose and treat illness in new ways. The most familiar medical radiation exposure is an x-ray. Even if you have never broken a bone, chances are you have still had an x-ray at the dentist at least a few times in your life.

The world of nuclear medicine encompasses a lot more than just x-rays and is utilized in ways you probably never would have guessed. This topic is much too large to fit in this book but some of the more notable procedures include CT scans, mammograms, brachytherapy, radioiodine therapy, and nuclear medicine used to kill cancerous cells.

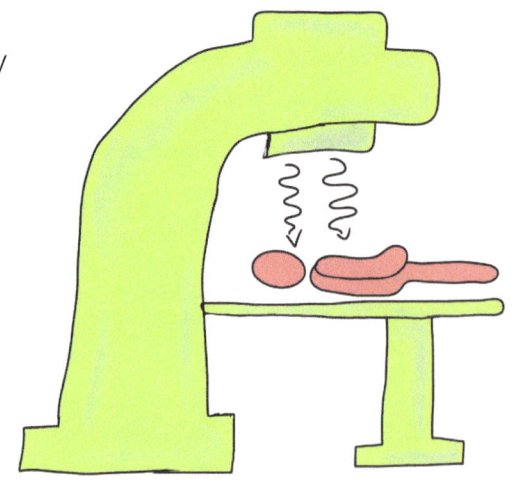

Types of Radiation

Radioactive decay comes in many shapes and forms. The type of decay an atom undergoes depends on what is causing the atom to be unstable. There is a fine balance that needs to be met with the number of protons, neutrons, and electrons; too many or few of one category can send an atom into a volatile state. The common forms of radioactive decay are alpha, beta, gamma, and nuclear fission.

Alpha radiation occurs when an unstable isotope emits two protons and two neutrons bound together. This alpha particle is identical to a Helium-4 nucleus. This is one of the largest particles that can occur during radioactive decay. Since both protons and neutrons are being released, both the element and isotope number will change. The daughter product will have an atomic number lower by 2, and an overall mass lower by 4 than the parent. Alpha decay tends to occur in very large atoms. Some examples of isotopes that undergo alpha decay include Uranium-238 and Platinum-175.

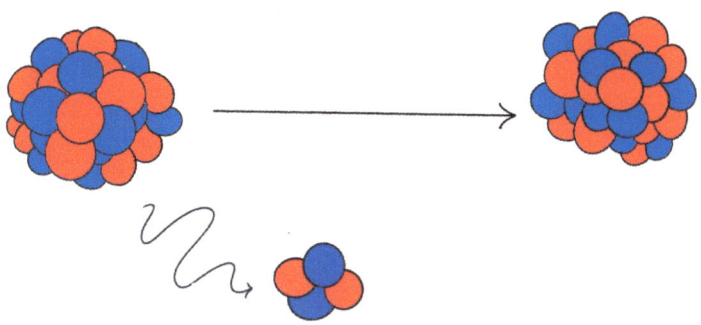

Beta radiation is broken into two different sub-categories. There is Beta Positive (b+) or Beta Negative (b-) decay. Beta decay occurs when the parent atom is unstable due to its charge. The imbalance in electric charge is because there are either too many protons or too many electrons. Beta decay allows the atom to move to an electrically neutral state. The decay pathway taken (b+ or b-) depends on whether there are too many protons (b+) or too many electrons (b-).

Beta Negative decay happens when an atom has too great a negative charge, meaning there is a surplus of electrons. In order to neutralize the charge, a neutron transforms into a proton, electron, and anti-neutrino.

Side Note: **Neutrinos** and **Anti-neutrinos** are subatomic particles with no charge. They are very abundant and do not cause significant interaction with matter.

The proton stays in the nucleus to balance the charge, while the new electron and anti-neutrino are ejected as radiation. The daughter product now has one more proton, so the atomic number increases and the element changes. There is no change to the weight (isotope number) because protons and neutrons have the same mass and we have converted 1 neutron into 1 proton. Examples of isotopes that undergo beta negative decay include Carbon-14 and Cesium-137.

Beta Positive decay happens when an atom is too positively charged due to an excess of protons in the nucleus. To become neutral, a proton breaks down into a neutron, positron, and neutrino. The positron and neutrino are emitted as radiation while the neutron stays inside the nucleus of the atom. Once again, the overall weight of the atom hasn't changed since a proton has been converted to a neutron. However, the number of protons has gone down and so the element is different. This time the atomic number decreased by one. Some examples of isotopes that undergo beta positive decay include Fluorine-18 and Magnesium-23.

> Side Note: **Positrons** are the opposite of an electron. They are the same size and weight as an electron but have a +1 charge instead of a -1 charge. Because of this they are sometimes called an anti-electron and fall in the category of anti-matter.

When an atom has undergone Beta Positive decay and released a positron there is a second type of interaction that occurs. Positrons can collide with electrons in a process known as **Annihilation**. Since positrons and electrons are opposite versions of the same particle, they combine and cancel each other out. All of the energy from the two particles becomes electro-magnetic waves called photons. Annihilation causes a positron and electron to become two photons that travel in opposite directions. This phenomenon can be utilized in procedures such as Positron Emission Tomography (P.E.T) scanning.

Gamma Radiation is made of electromagnetic waves called photons. Gamma rays (photons) are emitted when an atom is in an excited state and releases energy as it moves to a stable state called ground state. An atom is typically **excited** when an electron is in a higher orbit than it should be. The electron falls from the high orbit into its normal orbit closer to the nucleus to become stable. The energy of the electromagnetic wave released is determined by the distance between where the electron is and where the electron should be. A larger jump means more energy. There are no particles emitted in this radioactive decay only photons. The parent isotope and daughter product have the same atomic number and mass, the only difference is the configuration of the electrons. We denote the excited version of the atom by adding an asterisk * symbol. One example of an isotope that undergoes gamma decay is Plutonium-240* which decays to Plutonium-240.

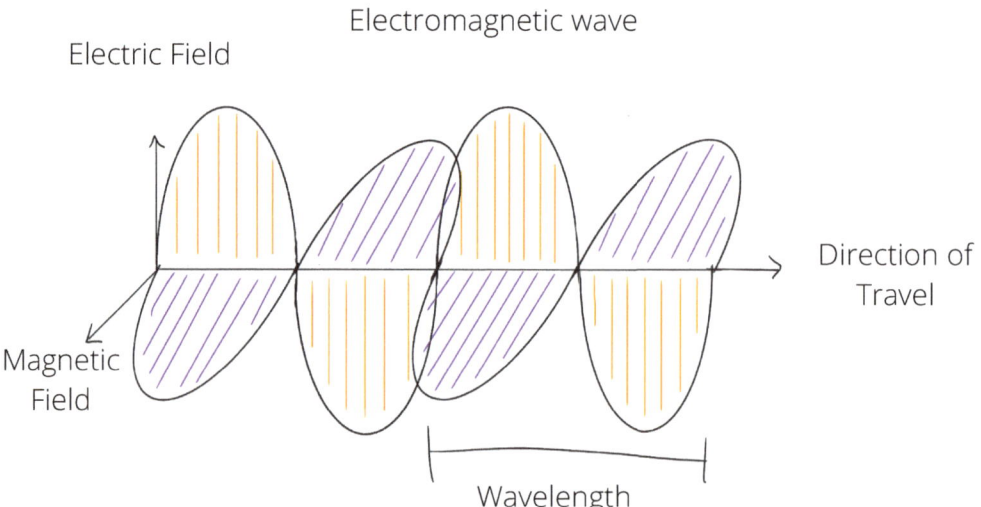

Gamma decay often happens as part of another radioactive decay method. When an unstable atom undergoes alpha or beta decay it can end up in an excited state because the arrangement of particles changes. A famous example of this is Cobalt-60 which is widely used for sterilizing medical tools. Cobalt-60 first undergoes beta negative decay which leaves it in an excited state. The excited atom then releases a gamma photon as it moves to the ground state.

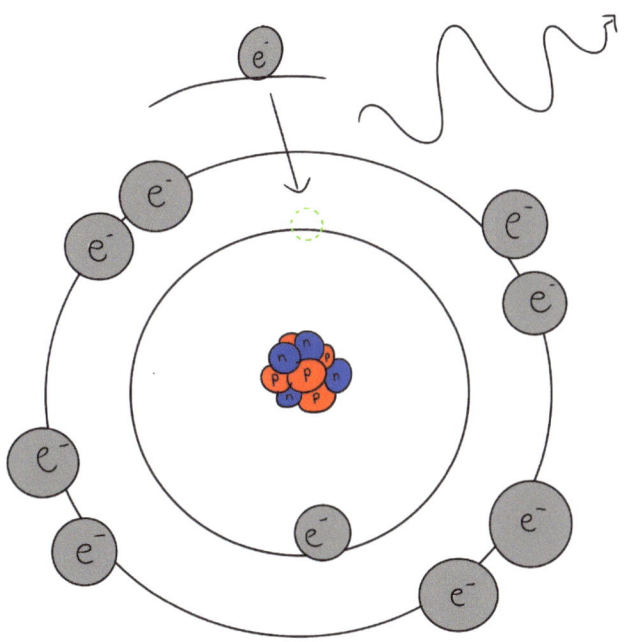

Neutron Radiation refers to neutrons that have been emitted, often during nuclear fission and fusion. Nuclear fission is when a large atom splits into two or more pieces, emitting extra neutrons in the process. Nuclear fusion is the combination of two smaller atoms into one large atom, again extra neutrons are emitted. Neutron radiation often causes a cascade of radioactive decay events. The neutrons will hit and combine with atoms they come into contact with, changing the isotope to a potentially unstable version. Due to their neutral charge, neutrons can penetrate extremely far into materials, making it difficult to shield. Shielding will be discussed in more detail in a later section. Uranium-233 is an example of an isotope that emits neutrons.

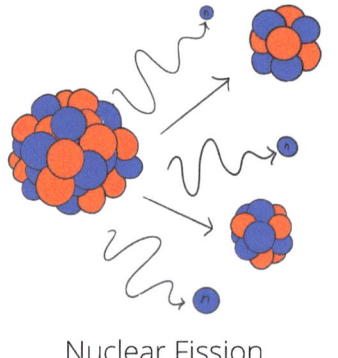

Nuclear Fission

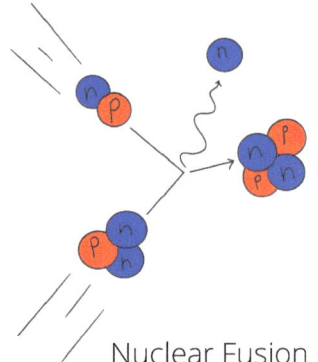

Nuclear Fusion

There are a number of other types of radiation, but those discussed here are the most prominent. Some less common radioactive decay methods include **internal capture** and **electron capture**. Since science is constantly evolving, we are likely to discover new forms of radiation in the coming decades. This is especially probable as we start exploring further into space and the solar system.

Dose Measurements

When people are exposed to radiation, it is important we have a way to characterize the dose received. Scientists think about **radiation dose** by the amount of energy absorbed into the body and how damaging it is. The specific dose measurement we care about most is called effective dose.

Effective dose lets scientists know how harmful the amount of radiation is. There are 3 factors that are taken into account for effective dose calculations. First is the amount of energy that was deposited into the body by the particles or electro-magnetic waves. Second is the kind of radiation (alpha particles, gamma rays, neutrons, etc) since different forms of radiation do more damage to cells than others. Third is the organ or tissue that was exposed. Some organs are very sensitive to radiation and damage easily, while others are not as affected.

Effective dose is measured in a unit called **Sieverts** which is given the short form Sv. One sievert is a pretty big dose of radiation, so it can be broken into smaller units. A microsievert (μSv) is 1/1,000,000 Sv and a millisievert (mSv) is 1/1000 Sv. Microsieverts and millisieverts are most commonly used to describe dose since most people only receive very small amounts. Think back to earlier when we talked about receiving an average annual dose of 2.4 mSv.

$$1 \text{ Sv} = 1{,}000{,}000 \text{ μSv}$$
$$1 \text{ Sv} = 1000 \text{ mSv}$$
$$1 \text{ mSv} = 1000 \text{ μSv}$$

A great way to put these units and doses into context is to compare them to something you are familiar with. My favourite way to do this is using bananas. Eating one banana causes approximately 0.1 μSv of dose due to the small amounts of radioactive potassium. We can therefore compare different situations where radiation dose is received, to the number of bananas you would need to eat to have the same dose. We call this unit the **Banana Equivalent Dose** and give it the short form BED. The following page has a fun chart comparing some exposure scenarios to bananas.

Number of Bananas	Equivalent Exposure
100,000,000	Fatal Dose (death within 2 weeks) (10 Sv)
20,000,000	Radiotherapy Session (2 Sv)
70,000	Chest CT scan (7 mSv)
20,000	Mammogram (2 mSv)
400	Flight from London to New York (40 μSv)
100	Average Daily Background Dose (10 μSv)
50	Dental X-ray (5 μSv)

Interactions with Matter

Radiation interacts with non-living matter in many ways. We call the material the radiation is interacting with the **target**. Each time radiation interacts with another atom, some of the energy is lost. The particle or wave will keep interacting until all the energy is deposited. The energy is transferred into the target material in a process known as **absorption**. The types of interactions depend on the kind of radiation, either charged particles (electrons, protons, and alpha particles) or neutral particles (photons and neutrons). Charged particles tend to have interactions with electrons of other atoms since these have a negative charge. The interaction could be a collision or a direction change due to an attractive/repellent force. This slows down the particle and causes energy to be transferred to the target atoms. Neutral particles tend to interact less with target materials since there are no electrostatic forces acting on them. In these cases, interactions only occur when the radiation hits something that is directly in its path.

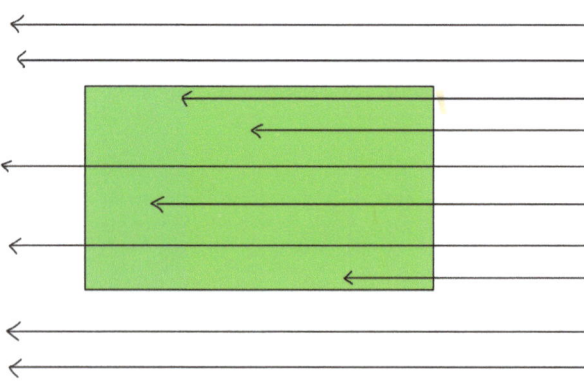

Radiation gets **scattered** as it passes through a target. The various interactions change the direction of the radiation like a ball in a pinball machine. Whenever the radiation hits something it gets deflected. These direction changes can be small, just a few degrees, or they can be drastic and send the radiation back in the direction it came from. The amount of energy lost depends on the direction change. A small direction change only costs a little energy, but a 180-degree turn will transfer a lot.

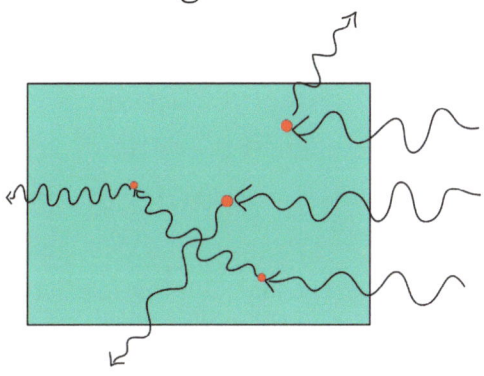

The biological damage caused by radiation comes down to how radiation interacts with the cells of living things. One of the main concerns that comes up is damage to the DNA of a cell. DNA is extremely important because it gives the cell instruction on how to grow, function, and replicate. Breaks in the DNA strand result in one of three outcomes: correctly repaired DNA, incorrectly repaired DNA (mutations), or cell death. Since DNA is made up of two opposite strands, a break in one strand is easy to repair by checking the opposite strand. However, if both strands are damaged or broken it is much more difficult to fix the damage, leading to mutations and cell death. This damage can occur directly or indirectly.

Direct damage is a result from radiation hitting strands of DNA. For example, incoming radiation in the form of fast electrons (like the ones released in beta negative decay) can directly hit a DNA strand and break apart the molecular bonds in the DNA backbone.

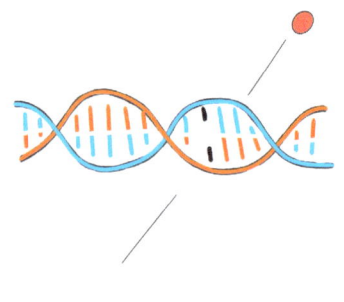

Larger radiation particles such as alpha particles are more likely to cause breaks in both strands causing worse biological effects. Since the size of a cell is very large compared to the DNA stored inside of it, the probability of a direct interaction is pretty small.

In most cases, the radiation causes **indirect damage** in the form of reactive oxygen species (ROS). Cells are mostly filled with water which can absorb the energy from radiation and break apart. The broken water atoms can become ROS and are very toxic. If there are only a few ROS they can float around the cell and come into contact with DNA causing breaks.

If there are many ROS inside of a cell it will cause so much damage to cell structures that it triggers cell death.

Side Note: **Reactive Oxygen Species (ROS)** are extremely reactive chemicals made up of oxygen and hydrogen that have additional negative charges. ROS include superoxide, peroxides, hydroxide ions, hydroxyl radicals and negatively charged water.

Dangers of Radiation

One of the most common questions that comes up when talking about radiation is *How much radiation is dangerous?* We are all used to getting x-rays at the hospital, UV rays from the sun, and radioactive potassium in our food. So at what point does radiation start to damage the body? The answer is not straight forward. There are two categories of negative effects we can consider, stochastic and deterministic.

Stochastic effects are genetic mutations that are caused by DNA that has been incorrectly repaired. The main stochastic effect we care about is cancer, but scientists also look at mutations that could be passed onto children. Stochastic effects can happen from any dose of radiation since even one damaged cell could cause that effect to occur. However, the chance that one damaged cell will cause cancer is extremely low. Due to this, stochastic effects are considered probabilistic. A higher dose of radiation causes more cells to be damaged and so the chances the effect will happen increases.

There is a probability of stochastic effects linked to each dose level. The probability of cancer is extremely low, even at the dose levels received by Nuclear Energy Workers, so it isn't something you have to worry about when you go to get your next dental x-ray. The lowest dose we have been able to observe any excess cancer risk is 100 mSv which is comparable to having more than 10 full-body CT scans done in a row or eating 1,000,000 bananas in a day.

Deterministic effects are more serious conditions such as radiation burns, cell damage, organ damage, acute radiation syndrome, and even death at extreme levels. Deterministic effects are caused by a very large dose received in a short amount of time. The severity of the effect also depends on the dose; a higher dose causes a worse effect. Think about sunburns, a form of radiation burn caused by UV rays. If you are only out for a short time you may only have a mild sunburn, but if you are in the sun for longer the burn gets worse.

Deterministic effects have a threshold dose, meaning a minimum dose must be met for the effect to occur. If you are below that dose, then the effect will not happen. The government regulates the radiation dose that workers and members of the public are exposed to in order to keep them safe. The dose limits put in place by the government are below the threshold for any deterministic effects to occur.

> Side Note: Most deterministic effect thresholds are measured in a unit called **Grays** (Gy) which only takes into account the energy deposited per kilogram of tissue. 1 Gy=1 J/kg. To convert these values into sieverts we have to take into account the type of radiation and where it has been deposited.

Burns are one of the most common deterministic effects seen. This is because most radiation sources are outside the body and the skin absorbs the majority of the dose. Radiation burns have a threshold of 3-5 Gy deposited into the skin. This correlates to 0.03-0.05 Sv if the dose comes from gamma rays. At 20 Gy deposited into the skin (0.2 Sv of gamma rays) the burn will cause blistering and peeling. At 50 Gy deposited into the skin (0.5 Sv of gamma rays) there is tissue death. Because the skin has many layers of dead cells above the active basal cells the effects of radiation can take days or even weeks to appear. The damaged layer won't be visible until all the layers above it have been shed off.

Acute Radiation Syndrome (ARS) is the most concerning potential outcome of radiation exposure and can be fatal. ARS occurs at very high doses delivered in short time periods and is caused by a significant amount of cell death. Doses this high have only been seen during extreme nuclear accidents such as the Chernobyl disaster. A patient with ARS will undergo several stages of sickness. Immediate effects can begin presenting a few minutes to a few hours after exposure. These effects include vomiting, diarrhea, headache, fever, and impaired cognitive function. The immediate effects stop after a day or two and the patient enters the latent period.

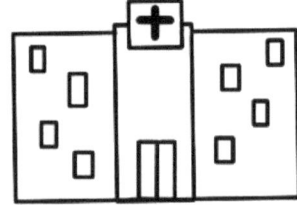

During the latent period, there are no negative symptoms, and the patient feels normal. The latent period can last several days or even a few weeks in very mild cases of ARS. Finally, the patient enters the illness period. Symptoms of the illness period depend on the severity of the exposure but can include decreased white blood cells, bleeding (internal or external), hair loss, vomiting, low blood pressure, shock, and more.

There are 4 levels of ARS. Each level can be fatal but the time between exposure and death varies. The levels of ARS let us know the most concerning symptoms/areas of damage. The lowest level is called **Hematopoietic** which is caused by damage to bone marrow cells. These cells produce various blood cells including white blood cells that help guard the body against infection. This form of ARS is dangerous because the immune system is susceptible, and a lack of blood platelets can cause excessive bleeding. Treatment is possible with a bone marrow transplant and blood transfusions. It can be very successful in mild cases if done soon after exposure.

The second level of ARS is caused by gastrointestinal damage. There is also damage to the bone marrow, but the GI tract damage is more serious and takes precedence. Without medical intervention it is very unlikely to survive this high of a radiation dose. Treatment includes immediate bone marrow transplants and symptom management.

The third level of ARS is characterized by damage to the cardiovascular system (heart and blood vessels) in addition to damage in the GI tract and bone marrow. Patients typically only live a few days and treatment is focused on keeping the individual comfortable during this time. A bone marrow transplant may be attempted but there is minimal chance of success due to significant damage in other areas of the body.

The fourth level of ARS includes damage directly to the nervous system (brain and nerves) as well as the heart, GI tract, and bone marrow. At this level there is no chance of survival and most patients do not live past 48 hours.

Level	Area of Concerning Damage	Range of Dose	Time Till Death	Survival Chance with Medical Care
1	Hematopoietic (bone marrow)	1-6 Gy	4-8 weeks	1-2 Gy >95% 2-6 Gy > 50%
2	Gastrointestinal Tract	6-20 Gy	10-20 days	< 50%
3	Cardiovascular System	20-50 Gy	1-5 days	0-1%
4	Nervous System	50+ Gy	<48 hours	0%

Radiation Protection

Since radiation can be dangerous it is important that we protect individuals who might be exposed. The ideal option would be to eliminate the radiation exposure entirely, but since that is not possible, radiation protection focuses on keeping the dose received as low as possible. Lowering the dose decreases the probability of stochastic effects like cancer and keeps the exposure below the threshold for a deterministic effect to occur. There are three major ways to protect someone from radiation exposure. **Time. Distance. Shielding.** Usually, all three of these methods are integrated into radiation protection.

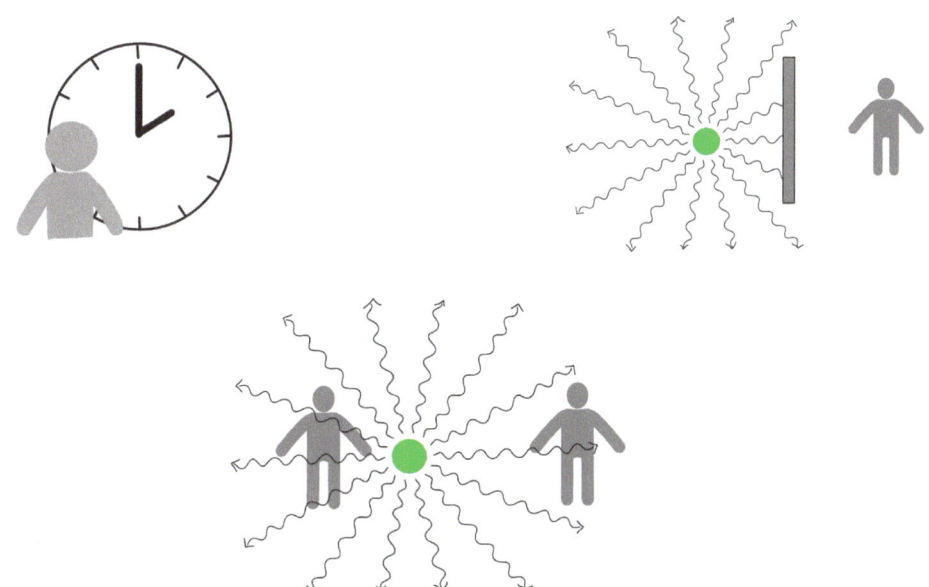

The first method to reduce radiation risk is to limit the time a person is exposed for. A shorter period near a radioactive source means less dose is received. Even though this method does not decrease the strength of the radioactive field, it can reduce the total dose.

For this reason, it is important that scientists only have their radioactive experiments out in the lab when being used. A worker or scientist never goes near a radioactive source unless there is a reason they need to. Similarly, in emergency situations involving radiation, workers will leave the area to make a plan on how to deal with things instead of pausing to think in a dangerous area.

Another way this method is integrated into radiation protection is in breaking up jobs. If work needs to be done in an area with a lot of radiation, often several workers will take turns so each individual person is only exposed for a fraction of the total job time.

The second method for reducing radiation dose is by standing farther away from the source. The strength of the radiation field decreases with distance, so having an extra meter or two makes a big difference. Think about fire, standing very close is extremely hot but if you stand back further the temperature is much lower. There is some math involved in understanding this fully, but the strength of radiation is proportional to $\frac{1}{r^2}$ where r is the distance between you and the radioactive material. So, if you were to double the space the strength of the radiation hitting you is ¼ as strong. There are a few different ways distance is implemented in the nuclear industry. The most prominent method for adding space between people and radioactive sources is by using tools. For example, most experiments are conducted using tongs to pick up samples or with pipettes to transfer liquids.

The third method is to add shielding between the radiation source and the individual. Shielding is a physical barrier that absorbs some of the radiation. Remember last time you got an x-ray done and you had to wear that heavy lead vest? That vest is an example of shielding being used. It was there to stop any x-ray exposure to body parts that didn't need to be imaged. Thicker shielding will block a larger percentage of the radiation while thinner shielding will let more pass through. There are many different materials that can be used to shield radiation and the one chosen depends on what kind of radiation needs to be blocked.

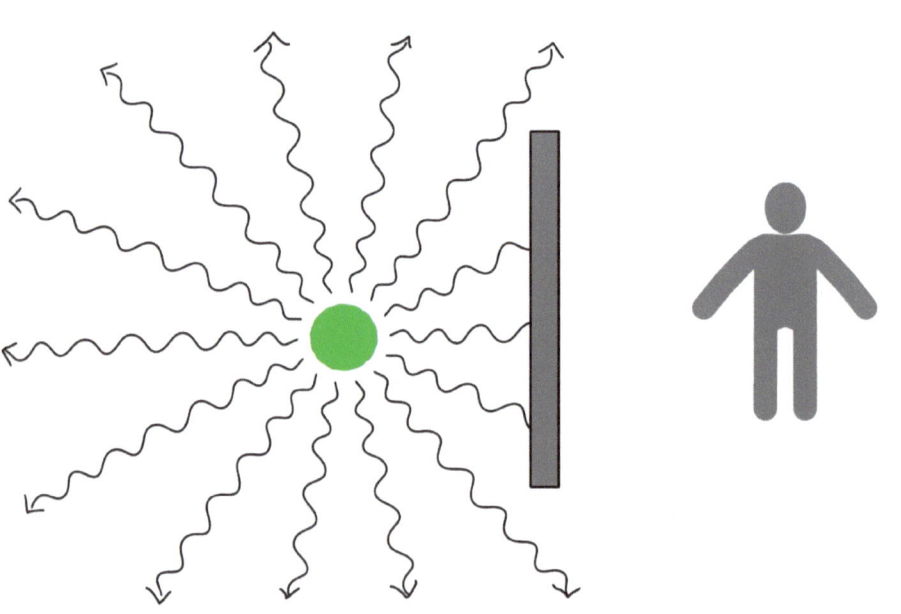

The hardest forms of radiation to shield are electromagnetic waves (gamma rays or x-rays) and neutrons. Gamma rays are best shielded by materials that have a high atomic number and are very dense. The most common element used for this is lead. Neutrons present a unique challenge when shielding since a neutron can be absorbed by whatever atom it comes into contact with. Absorbing a neutron increases the isotope number and can make the shielding material radioactive. Due to this, neutrons are best shielded by materials with a low atomic number like hydrogen, oxygen, and carbon. Isotopes of these elements tend to remain stable with an extra neutron. One of the most common ways to shield neutron radiation is using water. This is why many nuclear reactors are placed in a pool.

Where Does Radiation Come From?

There are a wide variety of sources that produce radiation. The most obvious are **natural sources**. These are radioactive elements that are unstable and undergo continuous radioactive decay. Elements like Uranium and Radium naturally have large percentages of unstable isotopes. Radioactive elements can be mined, refined from other elements, or produced as a daughter product in a decay chain. Many of the earliest applications of radiation relied solely on naturally occurring radioactive sources. In fact, many countries that do not have guaranteed access to electricity use only natural sources for medical imaging or treatment.

Due to the beneficial aspects of radiation, many artificial radiation sources have been created. One of the most common is an x-ray machine. An **x-ray tube** is able to produce electromagnetic waves, similar to gamma rays. A large current is passed through a wire causing electrons to be released. The negative electrons are attracted to a positively charged target nearby and accelerate towards it. As soon as the electrons hit this target they rapidly decelerate, and their energy converts to an electromagnetic wave. Some shielding is added around the path of the x-rays to absorb any radiation that is headed in an undesirable direction. Since the initial electrons causing this reaction are created by an electric current the machine can be turned on and off. X-ray machines are utilized for taking x-ray images or at high energies can be used for external beam radiation to kill cancer cells.

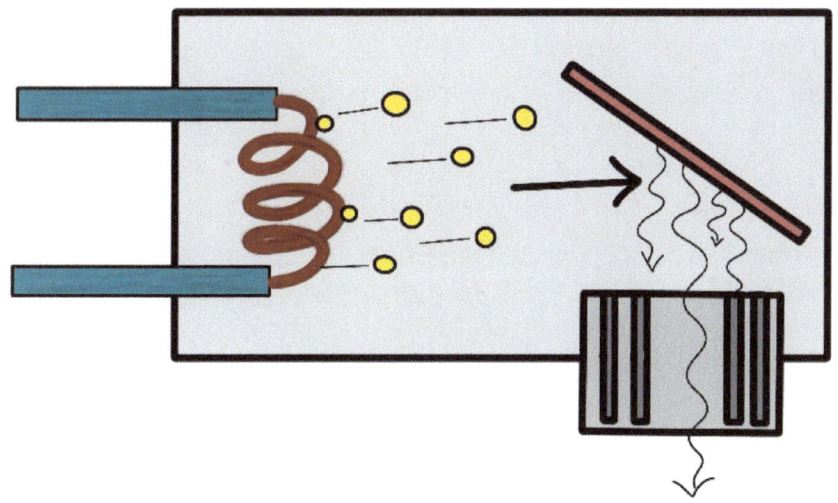

Cyclotrons are another type of particle accelerator mainly used for heavy particles like protons. Charged particles are accelerated by changing electrical currents inside the device. Walls of magnets bend the path of the particles causing them to move in a spiral. The spiral shape allows a much longer accelerating distance in a small space. The emerging particles exit the cyclotron as a high-energy beam. Cyclotrons are particularly useful for creating radioactive isotopes for medical use. In the case of a proton beam, the high-energy particles can interact with target atoms and change the element type of an atom.

Nuclear reactors harness the neutron radiation from nuclear fission. A nuclear reactor uses a natural radiation source as fuel, typically Uranium-235. The large atoms are unstable and break into 2 small atoms while releasing large amounts of energy and several neutrons. Some of the released neutrons hit other uranium atoms triggering a chain reaction. Nuclear reactors can be utilized for power generation or as neutron sources. The energy released through the fission reactions creates excessive amounts of heat that can be captured by a steam turbine. Nuclear reactors are one of the most efficient ways to produce electricity and make up a significant portion of the power supply in North America. The excess neutrons can also be used for the production of isotopes. Neutrons hitting a target element will be captured and increase the isotope number of an atom.

Decay Kinetics

Radioactive decay alters the parent atom causing it to become the daughter product. This makes radioactivity hard to characterize since over time the source of radiation will change. In the beginning, there will be many parent atoms that are undergoing decay but later on, there are fewer parent atoms left. Time is a very important factor to consider when thinking about radiation since it has a big impact on how radioactive something still is. Looking at radiation in the context of time is called **decay kinetics**.

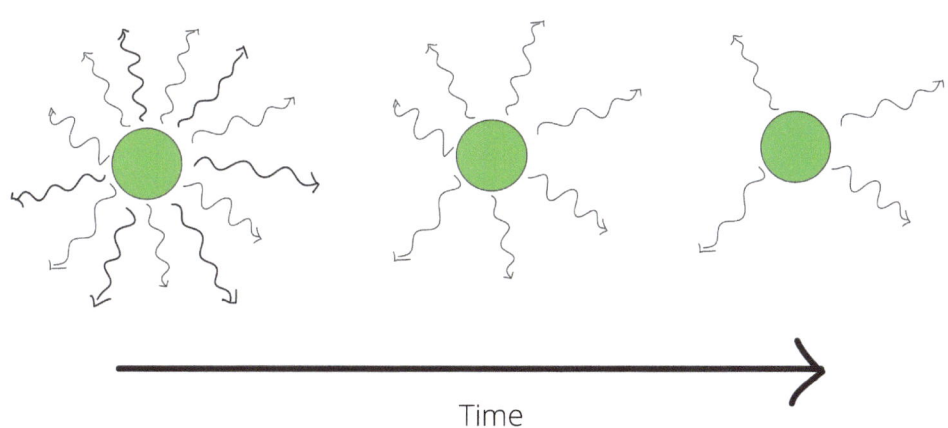

Time

Some Isotopes take longer than others to undergo radioactive decay. The time it takes an isotope to decay can be nanoseconds long or it could be millions of years. The rate of decay is unique to each isotope and we can characterize this in a quality called **half-life**. The half-life of an isotope is the time it takes for half of the atoms to undergo radioactive decay. The rate of decay is not linear, it actually follows an exponential decay curve. After one half-life 50% of the original amount remains, after 2 half-lives 25% remains, then 12.5%, and so on until virtually all of the isotope has decayed. A radiation source will be very radioactive during its first half-life but gets weaker and weaker over time. Isotopes that are very unstable tend to decay faster, meaning they have a shorter half-life. Similarly, atoms that are only somewhat unstable will decay much slower and have a longer half-life.

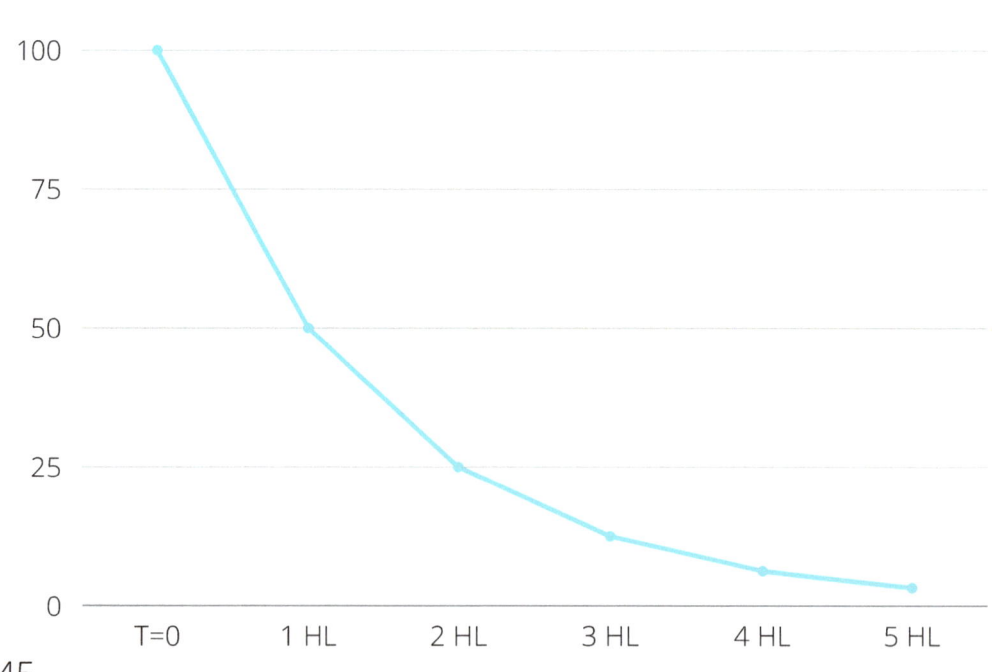

The measurement we use to describe how many radioactive particles or electromagnetic waves are coming from a source is called **activity**. We can define activity as the number of radioactive decay events per unit time. The most common unit used to measure activity is **becquerels** which is defined as 1 decay per second. We describe activity by saying hot, cold, and dead. A "hot" source is very active and is giving off a lot of radiation. A "cold" source is barely active at all and only undergoes a few radioactive decays per second. Finally, in a "dead" source, all the parent atoms have decayed to daughter products. There is no more radiation coming off the source and no decay events happening. The activity of a source changes over time but we can use formulas to calculate this. All we need to know is the amount of activity when we started, the half-life of the isotope, and how much time has elapsed.

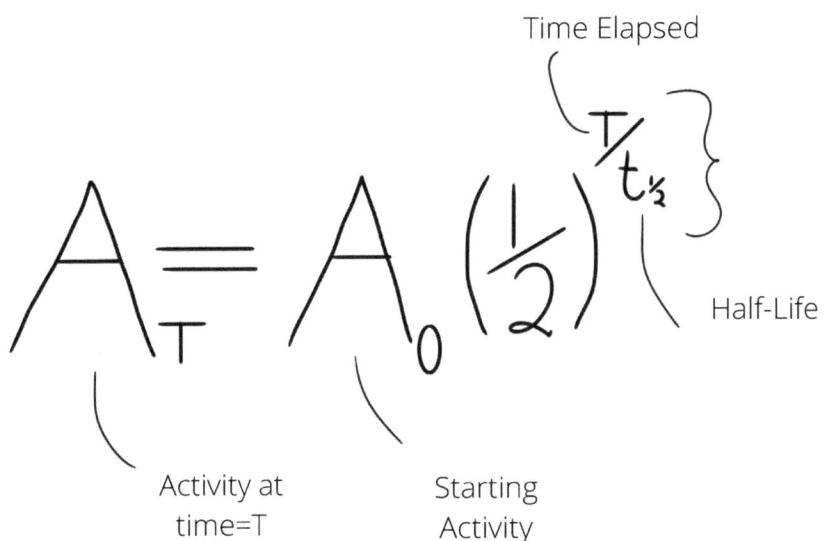

$$A_T = A_0 \left(\frac{1}{2}\right)^{T/t_{1/2}}$$

- A_T: Activity at time=T
- A_0: Starting Activity
- T: Time Elapsed
- $t_{1/2}$: Half-Life

How X-Rays Work

X-rays are the form of radiation we come across the most in medical procedures, but how do they actually let us see inside a person? The reason x-rays are useful for imaging bones comes down to the different elements that make up bones and soft tissue. Soft tissue such as muscle and fat are made of elements with low 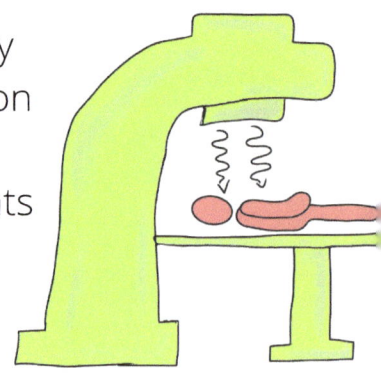 atomic numbers. The main elements present are carbon, hydrogen, oxygen and nitrogen. Bone contains elements such as calcium and phosphorous, which have higher atomic numbers. X-rays are absorbed differently  depending on the elements they are passing through. Elements with high atomic numbers absorb more x-rays than those with low atomic numbers. Due to this, many x-rays pass through soft tissue, but bones block the x-rays from getting through. A detector on the other side of your body counts the number of x-rays that hit it.

The detector is set up like a grid where each square becomes a pixel of the final image. Squares where many x-rays hit show up dark while areas that had only a few x-rays show up white, giving us a look inside your body.

Cool Uses for Radiation

There are a lot of cool uses for radiation you may not know about. The medical sector uses radiation for a lot more than just x-rays and killing cancer cells. Radiotracers are a very useful way to understand how the body is functioning. Various radioactive elements can be attached to molecules and sent into the body. The radioactive tracer lets us measure and track where these molecules travel. Some of these tests look at the function of organs like the thyroid, liver, kidneys, heart and lungs. Other tests look for cancers by attaching radiotracers to glucose molecules which accumulate in tumours. One very common radiotracer is Technesium-99.

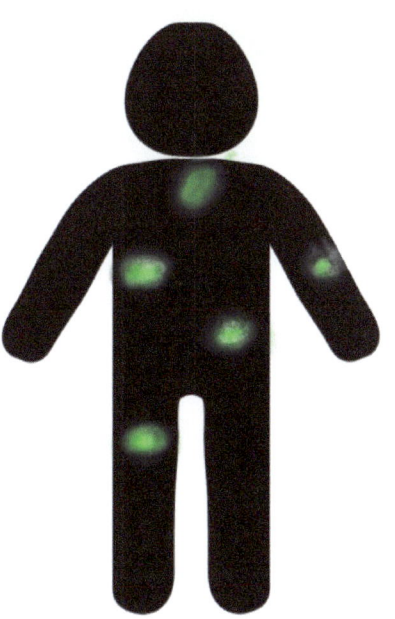

This element works particularly well because it has a half-life of 6 hours and emits a gamma ray strong enough to travel out of the body and to a detector. The 6-hour half-life gives doctors plenty of time to take images but by the next day there is very little radioactivity left inside the patient.

Strong gamma sources like Cobalt-60 are often used for sterilization. The radiation field is damaging to any microorganisms that may be present and all bacteria are killed in the process. Items only need to be placed in the radiation field for a few minutes to completely sterilize the material. Since gamma rays are very penetrating, they can pass through the entire object creating a uniform cleaning. This is commonly used for single-use medical equipment like surgical instruments, gowns, masks, catheters, implants, and more. Gamma sterilization is also used for foods. Sterilizing foods can help reduce foodborne illnesses by killing any organisms inside like E. coli. The radiation also kills organisms that cause food to decompose, extending the shelf life.

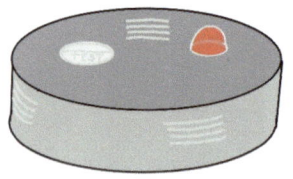

The smoke detectors in your house make use of a radioactive source as well. Most smoke detectors use Americium-241 to detect changes in the air. Americium-241 ionizes the air around it when it undergoes alpha decay. This produces a small consistent electrical current. Smoke particles stick to the ions created by Amermicium-241 and disrupt the current triggering the alarm to sound.

Radiation can help us figure out how old something is through a process called radiocarbon dating. This process uses the ratio of Carbon-12 and Carbon-14 present in the artifact. All living things absorb carbon through the air or in food. A portion of this will be in the form of Carbon-14.

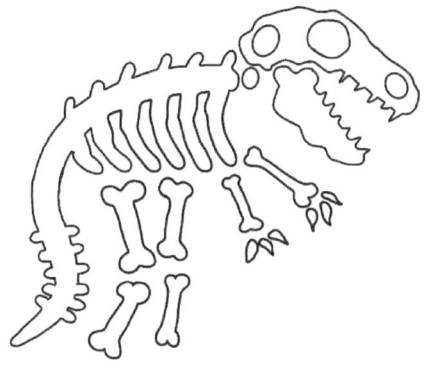

However, when the plant or animal dies it no longer takes in new Carbon-14 from the environment around it. The unstable Carbon-14 decays to a more stable isotope. By analyzing how much of the Carbon-14 has decayed or not decayed, researchers can measure how old something is.

Radiation can also be harnessed to determine the elemental composition of a sample. Neutron Activation Analysis (NAA) uses neutrons to create unstable isotopes which undergo radioactive decay. When an atom absorbs a neutron, the new isotope may be unstable and quickly decays back to a more stable form. The energy released is unique to the element that has been excited. By measuring a spectrum of the energies emitted, researchers can determine the elements present in a sample and the ratios of each.

A Brief History of Radiation

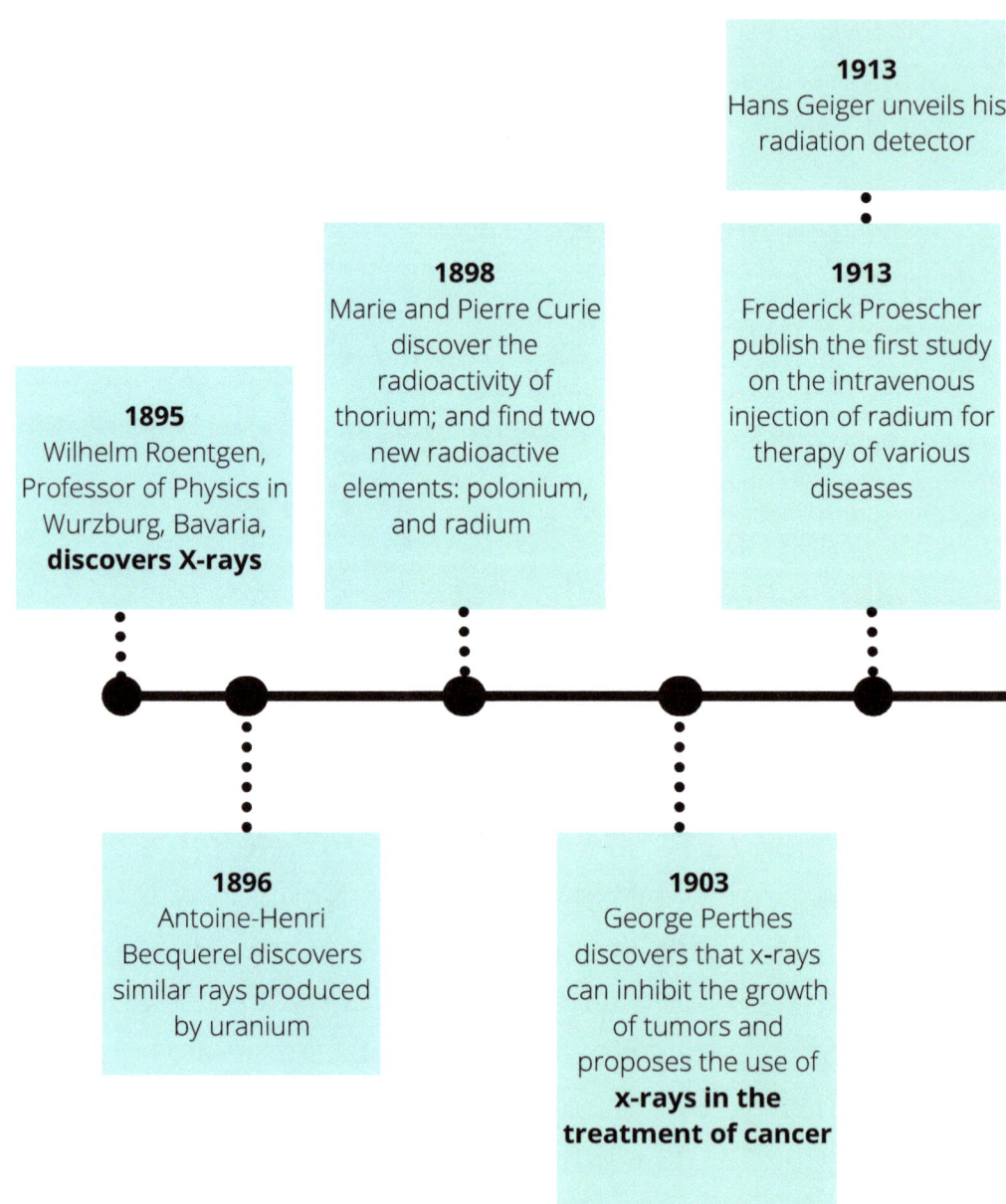

1895 Wilhelm Roentgen, Professor of Physics in Wurzburg, Bavaria, **discovers X-rays**

1896 Antoine-Henri Becquerel discovers similar rays produced by uranium

1898 Marie and Pierre Curie discover the radioactivity of thorium; and find two new radioactive elements: polonium, and radium

1903 George Perthes discovers that x-rays can inhibit the growth of tumors and proposes the use of **x-rays in the treatment of cancer**

1913 Hans Geiger unveils his radiation detector

1913 Frederick Proescher publish the first study on the intravenous injection of radium for therapy of various diseases

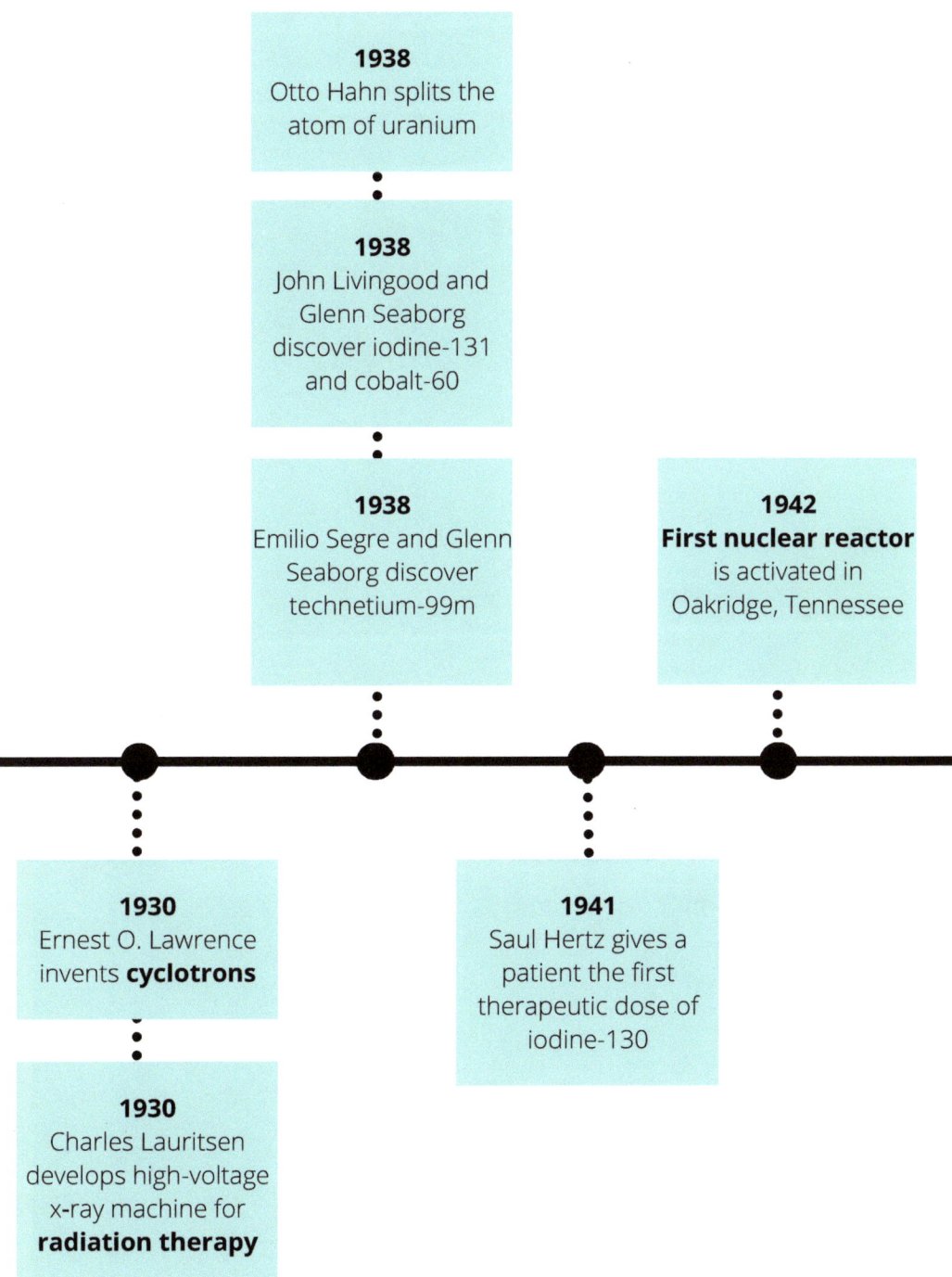

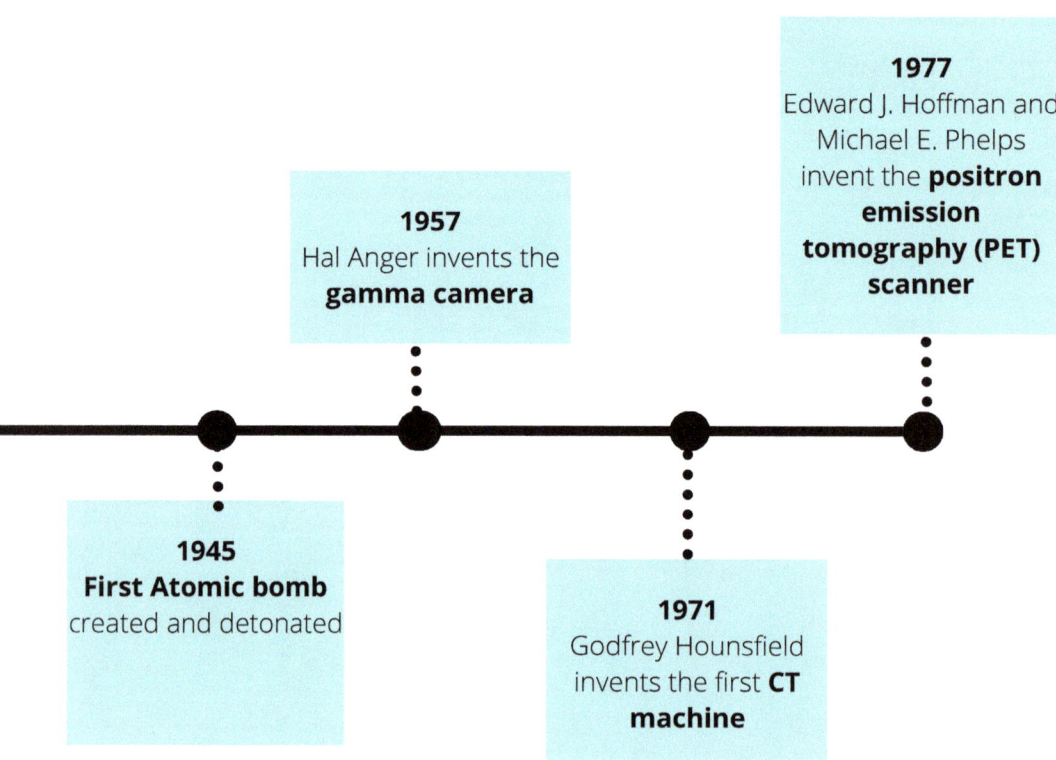

www.ingramcontent.com/pod-product-compliance
Lightning Source LLC
Chambersburg PA
CBHW040240220526
45473CB00001B/312